MARIJUANA
MULTIPLIER

Notice to Reader

Dr. Atomic's Marijuana Multiplier is adult literature and not for children and is made available for informational, archival, and entertainment purposes, in accord with the First Amendment. First published in the early 1970s, the cartoons chronicle the underground and counterculture flavor of the period. *Dr. Atomic's Marijuana Multiplier* has been republished as a historical novelty and archive of underground processes utilized at the time. *Please note that the author and publisher advocate no illegal activities of any kind, and make no express or implied warranties of merchantability or fitness for any purpose, with respect to this book or the information it contains.* Cautions have been added to this revised edition to remind the reader that the processes described are presented for informational purposes and should be carried out by trained experts under controlled conditions only.

MARIJUANA
MULTIPLIER

Larry S. Todd

RONIN PUBLISHING, INC.
BERKELEY, CA
www.roninpub.com

DR. ATOMIC'S MARIJUANA MULTIPLIER
ISBN: 1-57951-003-5
ISBN: 978-1-57951-003-9
Copyright: 1974 by Larry S. Todd (First Ed)
Copyright: 1998 by Ronin Publishing (Derivative)

Published by
RONIN PUBLISHING INC
PO Box 3436
Oakland CA 94609
www.roninpub.com

Derivative Author:	Beverly Potter
Project Editor:	Sebastian Orfali
Page Composition:	Beverly Potter
Cover Design:	Generic Type
Drawings:	Larry Todd

Published in the United States of America
Distributed by PGW

Dr. Atomic's

Marijuana Multiplier

Greetings!

I'm your host,

Dr. Atomic.

Across the table from me you'll see our guest, the plant
Cannabis Sativa!

Today we'll be reviewing methods of making Cannabis *two, three, or more times stronger* ! We'll cover procedures for making hash, hash oil, isomerization of cannabidiol—
a chemical process which creates more THC, one of marijuana's most important ingredients—and sundry other items of arcane pharmacopoeia.

There are several methods for increasing potency of marijuana. The resulting preparations can be smoked or used in marijuana foods. The greatest advantage, however, of increasing the potency of marijuana is in smoking, as opposed to eating the marijuana. Increasing the potency of marijuana means that a much smaller amount can be smoked to achieve the desired effects. Therefore, increasing potency minimizes the health hazards and physical irritations associated with smoking.

CAUTION: Processes described in this book involve volatile solvents and Sulfuric Acid which are dangerous chemicals. Anyone interested in the methods described in this book should work under the supervision of a qualified chemist in a licensed laboratory and pay close attention to the cautions included.

Preparing the Marijuana

Supplies:
Raw marijuana or hashish
Blender

The plant material to be used must be processed for cleanliness. When marijuana is used, the seeds are first removed. The remaining flowering tops, leaves, and stems are broken down into bits as small as possible. The best flowering tops can be left aside for a process that involves them.

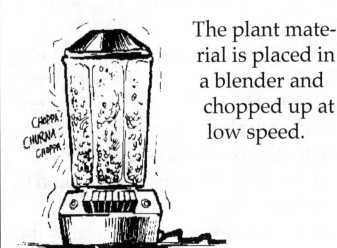

The plant material is placed in a blender and chopped up at low speed.

USING HASH

When hashish is used, it broken up into small granules before being put into  the blender since hashish can be dense and may jam or stall the blades of a blender. If the hash is hard, it can be cut with a sharp knife that has been heated in a gas flame to a dull red glow. Hash can be shredded on a cheese grater, or heated *gently* in an oven until it begins to soften. Powdered hash should not be left exposed, as it will quickly lose its potency.

Refluxing

The next step is *refluxing,* using a solvent to extract the vital resins from the matrix material. Once all of the vital principle has been extracted, the solvent is removed and the resins are isolated. The refluxing apparatus that Dr. Atomic uses can be built from common kitchen utensils and supplies available from most hardware stores.

ONE INCH WIDE STRIPS OF RUBBER INNER TUBE

ICE

LARGE POT LID

POLYETHYLENE TRASH BAGS

LARGE STEW POT

STAINLESS STEEL POT

ONE INCH THICK ROPE SPACERS

SOLVENT WITH MARIJUANA

TWIN PLATE HOTPLATE

Using Solvents

The marijuana is placed in a solvent for refluxing, and solvents are used at several later stages in the multiplier process.

CAUTION: Volatile solvents are hazardous, potentially explosive and toxic, and should be used only by professional chemists working in licensed laboratories. Solvents should never be used in the presence of motors, engines, or other devices that can shoot sparks, including cigarette lighters and ovens with pilot lights. Cigarettes, including joints, should never be smoked in the presence of these solvents, either.

Now a word on several solvents that can be used in the reflux apparatus, along with their pros and cons!

Methanol or ***Methyl Alcohol,*** also called wood alcohol, is available at some pharmacies and camping goods stores, where it is sold as fuel for camp stoves. Care must be taken to ensure that the methanol is pure (the ingredients list on the container can be checked), since it is occasionally found with other substances mixed in.

NOTE: Methanol boils at 149 degrees F which is 65 degrees C.

CAUTION: Methanol is quite poisonous and its toxic fumes are explosive. When used it must be kept totally sealed in the refluxing apparatus, and must be completely removed from the final product. How to do this will be described later.

Ethanol, also known as *Ethyl Alcohol,* or *grain alcohol,* is an excellent solvent. It is the active principle in alcoholic beverages, so it is heavily taxed and rather difficult to procure in pure form. "Denatured" spirits contain toxic substances intended to render the stuff unfit for drinking. Fortunately these evaporate at the same temperature as the alcohol, so they provide little difficulty in use as a solvent, so long as there is no solvent remaining in the extract.

NOTE: Ethanol boils at 173 degrees F which is 78.5 degrees C.

Isopropyl or *Rubbing Alcohol* is another solvent available at drugstores, however it is frequently adulterated with water—

usually about 30 percent. The presence of the water results in the extraction of water-soluble tars that give the product a bad flavor when smoked.

If, however, the oil is later re-extracted with a solvent without water in it, the water and tars it brings along can be removed easily.

NOTE: Isopropyl alcohol boils at 180 degrees F which is 82 degrees C.. Water boils at 212 degrees F
or 100 degrees C.

Petroleum Ether used as a solvent produces a smaller yield of extract than alcohol solvents discussed.
Its advantage is that the resulting extract is of greater potency by weight than that yielded by alcohol. Its greatest disadvantage is that, like gasoline, petroleum ether is explosive and therefore hazardous to work with.

CAUTION: Petroleum ether is toxic and highly explosive. Therefore, the less used, the better. Performing a reflux process with Isopropyl or Ethyl Alcohol before refluxing with petroleum ether allows less petroleum ether to be used. Since petroleum ether is a distillate product, its boiling temperature depends on what temperature and pressure it was distilled at. This means that petroleum ethers boil over a range of temperatures, usually 86 -140 degrees F, which translates into 30 - 60 degrees C.

IMPORTANT: Petroleum ether *must* be stored in a freezer and *must* be ice-cold when poured because some kinds boil at room temperature, releasing toxic and explosive fumes.

Making a Reflux Apparatus

Supplies:
Selected solvent
1 inch-wide strips of rubber inner tube
Lots of ice
Polyethylene trash bag
Twin-plate hot plate
Metal tub to hold twin-plate hot plate
Large stew pot
Stainless steel pot to fit into stew pot
Stainless steel wok without handles
        or pot lid too big to fit stew pot
1 inch rope

To start, the metal tub is positioned
securely on the hot plate and strips
of one-inch thick rope are placed
across the bottom of the tub. Next
the large stew pot is placed atop the
ropes. Mashed marijuana is placed in
a stainless steel pot. The stainless steel
pot is then placed inside the stew pot,
with the solvent containing the mari-

juana mash inside it. Enough solvent is poured in to immerse the marijuana. Then enough solvent is added so that the volume of marijuana mash plus solvent is roughly twice the original volume of the marijuana by itself.

A stainless steel wok with the handles removed or an oversized pot lid is placed upside-down over the large stew pot. Next a polyethylene trash bag is laid over the wok or oversized pot lid and fastened tightly to the sides of the large stew pot with one-inch strips of rubber inner tube.

Air that has become trapped under the garbage bag is forced out with the hands. A large amount of ice is piled on the plastic bag over the wok or oversized pot lid and the tub is filled halfway with water.

First Reflux

The hot plate is turned on until the water boils which heats the apparatus to about 212 degrees F or 100 degrees C. As the pot containing the marijuana and solvent is heated, the solvent boils and rises within the large stew pot until it contacts the chilled bottom of the wok or oversize pot lid, where it condenses. Dripping off the curved bottom of the wok or the handle of the pot lid, the solvent falls back into the stainless steel pot with the mashed medicine inside it. The mixture is kept in the smaller pot because the condensing area must be larger than the boiling surface.

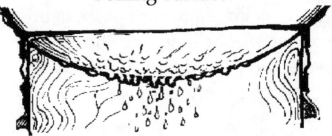

Plastic sheeting is used for several reasons. First, it completely seals off the atmosphere inside the reflux apparatus from the external atmosphere, preventing the leakage of toxic fumes.

Second, if the plastic begins to inflate, it is a sign to the chemist that it is starting get hot inside, pressure is building up, and more ice should be placed in the wok or on the oversized pot lid.

The marijuana mash is refluxed for three to four hours which is enough time to remove all the resins from the plant material.

CAUTION: The reflux operation must be watched closely so that ice can be added whenever necessary as indicated by inflation of the garbage bag.
It must never be left unattended. Pressure can build up inside the apparatus and cause explosions, fires, or release toxic fumes.

After refluxing, the hot plate is turned off and the apparatus allowed to cool.

Cooling the Assembly

Supplies:
Ice
Water
Tub
Blanket

The reflux apparatus is cooled by setting it into a tub of ice and water.

CAUTION: A blanket should be kept soaking in this tub to be used as an extinguisher in case trouble develops.

Reflux with Filter

Supplies:
Holding vessel (jar or pot)
Colander
Large piece of filter paper
Strainer

At this point, the stainless steel pot contains a soupy mixture of mashed marijuana cellulose and solvent with resins in it. Next this material is transferred into a holding vessel (a jar or pot of appropriate size), the emptied stainless steel pot returned to its position in the reflux apparatus, and a colander is placed over its top.

A large piece of filter paper is placed in the colander, and a strainer is placed on top of it. The solvent and mash are poured into the strainer to filter through the paper and colander and back into the stainless steel pot.

The mash—consisting of stems and cellulose which still contain a fair amount of resins—gets caught in the strainer and the filter paper below it where it remains during the second reflux, while the resin-bearing solvent collects at the bottom of the stainless pot.

The apparatus is set up as before. It is placed in the tub atop the twin plate hot-plate, with the wok or oversize pot lid at the top of the assembly, and sealed carefully with a poly-ethylene trash bag upon which a large amount of ice is then placed. The water in the tub may need replenishing as some of it will have boiled away.

Next, the hot-plate is then turned back on. As the syrup boils, the solvent fumes condense on the bottom of the wok (or around the handle area of the oversize pot lid), and then drip down through the strainer, mash, filter paper, and colander—extracting resins along the way—to collect back in the bottom of the steel pot.

STRAINER

FILTER PAPER

COLANDER

An hour is usually enough time for a second reflux to extract a maximum amount of resins from the mash. The cooling step is repeated.
CAUTION: A close eye must be kept on the reflux at all times. It must never be left unattended.

Removing Solvents

Supplies:
Round cake pan wider than steel pot
Container to store re-purified solvent
Colander

When the reflux assembly is cooled, solvents are removed from the extract collected in the stainless steel pot.

The filter and the strainer containing the leached marijuana plant matter are removed and put away to dry. This material can be used to make hash in a process that will be explained later.

A round cake pan, somewhat wider than the stainless steel pot, is placed in the colander.

The colander, with the cake pan in it but without the filter and strainer used in the last process, is now placed on top of the stainless steel pot, which still holds the syrup of solvent and resins. The stainless steel pot with the colander and cake pan on top are placed back inside the large stew pot, and the assembly is carefully sealed as before and topped with ice.

Water is replenished in the tub, as necessary, and the apparatus is heated again. Solvent will collect in the cakepan, rather than running back into the steel pot as before. An hour or two of this operation will produce a cake pan holding essentially pure solvent and a stainless steel pot now containing a material known as *crude marijuana extract.* The purified solvent that has collected in the cake pan can be stored in a tightly closed  container kept at low temperature for later use.

Removing Toxins and Water From The Extract

Supplies:
Pot large enough to put the stainless
　　steel pot inside
Candy or deep-fry thermometer
Small hand-held mirror
Cooking oil

The crude marijuana extract should now be largely free of solvent, but some residual amount of the solvent that has been used is still present.

CAUTION: Methanol and petroleum ethers are very toxic. Isopropyl alcohol may or may not present problems if present in small amounts in a smoking preparation, but it *is* very toxic when eaten, and therefore must be prevented from finding its way into cannabis foods. By itself, ethanol—which, after all is the active ingredient responsible for the alcohol "buzz" generated by liquors—isn't prohibitively toxic in the small quantities that would be left if the cake pan procedure was carried out correctly, but ethanol solvent products usually have highly toxic chemicals added in to deter any temptation to try to drink the stuff.
For all of these reasons, chemists perform a procedure to purify the crude

marijuana extract of possible chemical residues. Even if there aren't significant amounts of toxins left, the procedure for removing them is still important because it also removes any water left in the extract. Too much left-over water makes the end-product less cohesive and harder to smoke because the dampness makes it harder to ignite.

The first step in removing chemical residues and left-over water is to get a large pot—big enough to hold the stainless steel pot inside it—and fill it with a cooking oil like canola, corn oil, peanut oil, or olive oil. Then, the stainless steel pot with the extract is put inside it. A candy thermometer or a deep-fry thermometer is clipped to the rim of the big pot, with the mercury reservoir at the bottom of the thermometer submerged in the oil.

THERMOMETER

STAINLESS STEAL POT
CONTAINING EXTRACT
& WATER

COOKING OIL

Cooking Oil Assembly

The big pot and its contents are placed put on the hot plate. A small amount of water is put in the extract. The hot plate or oven burner is then turned on, and the cooking oil is heated until the thermometer reads about 220 degrees F or 104 degrees C. When a small mirror no longer fogs when it is held over the stainless steel pot, all water and toxic chemicals will have been driven off because they evaporate at a lower temperature than water.

This procedure produces a material that is called *Purified Crude Marijuana Extract.* Purified Crude Marijuana Extract can be smoked or  used in food preparations. It's a far stronger medicine by weight than the marijuana it was made from, but it still has lots of non- ingredients that can be removed from the resin by the exciting procedure I'll describe next!

Re-Dissolving the Extract

Supplies:
Petroleum ether
Water
Large bottle, jug, or jar
   with screw top

The first step in refining the Purified Crude Marijuana Extract is to dissolve it in five times its weight in petroleum ether in a large bottle, jug, or jar that comes with a tight-fitting screw top. A volume of water equivalent to that of the petroleum ether is then added.

CAUTION: Due to explosiveness of the petroleum ether, the mixture must be kept cool.

A volume of petroleum ether equal to half the volume of water used is added. The cap is tightened and the jar is turned upside down and then turned back upright immediately. The mixture is allowed to run down the sides of the jar for a moment. The jar is uncapped to relieve pressure, and recapped. This procedure of quickly turning the bottle upside down and back upright again and briefly uncapping it is repeated about 25 times.

The jug must sit for about a half-hour to let the contents to settle and separate into three distinct layers. At the bottom will be a layer of alcohol, water, tar, and spare resins. The middle layer will be a dark emulsion layer, and the top layer will consist of petroleum ether extract.

Blowing Off The Petroleum Ether Extract Layer

Supplies:
2 large glass jugs
2 pieces of rubber tubing
2 2-hole rubber stoppers
4 glass tubes, 1 long and 3 short
Fresh petroleum ether

Supplies are assembled into the blow-tube set-up pictured on the next page. The resulting apparatus is used to blow the petroleum ether-extract layer off the top into a collection jar.

The chemist gently blows into the jug until the petroleum ether-extract layer at the top is siphoned off into the collection jar. Drawing off any of the emulsion layer at the bottom is avoided while the remaining mixture is allowed to settle into three layers.

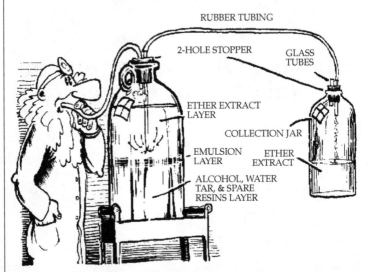

A volume of fresh petroleum ether equivalent to the amount used in re-dissolving the extract is added, and the blowing-off procedure is repeated as many times as necessary for the petroleum ether-extract layer to be clear after settling.

Refluxing the Petroleum Ether Extract Solution

Supplies:
Reflux set-up with cake pan
 and colander
Boiling water bath

The petroleum ether-extract solution in the collection jar is poured into the stainless steel pot. The colander and cake pan are positioned on top of the stainless steel pot, which is put inside the refluxing appa- ratus, which is then charged up with ice as before and slowly heat- ed to about 140 degrees F or 60 degrees C.

After the petroleum ether is evaporated it can be saved for later use. The pot containing the oil is placed in a bath of boiling water for a few moments to remove any residual traces of petroleum ether.

The yield produced is much smaller than in the initial steps, but the potency is far greater. The material that has been made is called Purified Marijuana Extract, as distinct from Purified *Crude* Marijuana Extract.

Isomerizing

Potency of the Purified Marijuana

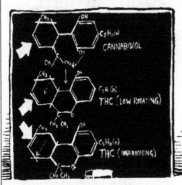

Extract can be increased even more by a process called *isomerizing*. In this process, a molecule is converted into an *isomer* of itself—a version that has the same atomic contents but a different structure which has different properties. The isomerization procedure I will describe converts the THC present in the extract into an isomer of itself which is far more potent! This is accomplished by altering the "low-rotating" forms of THC present into higher-rotating isomers. "Rotation" refers to the double molecular bonds indicated in the diagram and their position on the left-hand carbon ring. In low-rotating THC, the double molecular bond is in a lower position. Isomerization moves it to a

higher position. This procedure simultaneously turns cannabidiol, another ingredient in marijuana, into THC. Since cannabidiol isn't psychoactive like THC, this makes the extract more potent in terms of its psychoactive effects. However, cannabidiol has appetite-stimulating, pain-reducing, anti-inflammatory, and other effects of its own, separate from those of THC. Some marijuana users may find that they prefer not to isomerize their extract in order to retain the effects of cannabidiol, which will be lost through this isomerization process. Medical marijuana users seeking psychoactive, anti-spasmodic, and other effects of THC will find this isomerization process beneficial.

The psychoactive potency of extract which has been processed in this manner may at least be doubled, though in some cases it may be as much as five or six times stronger!

CAUTION: This method involves the use of pure, concentrated Sulfuric Acid, a dangerous reagent, so qualfied chemist use protective clothing to protect themselves.

Isomerizing with Sulfuric Acid

Supplies:
Pure ethanol or methanol
Sulfuric acid
Beaker or vessel for dissolving
Long glass stirring rod
Heat proof glass pot
Refluxing apparatus
Petroleum ether
Petroleum ether extraction apparatus
  with blow tube

Sodium bicarbonate (baking soda)
Safety glasses
Long rubber gloves
Rubber lab apron
Clothing that covers as much of the
              skin as possible
Face mask
Sodium bicarbonate (baking soda)
   solution
Styrofoam- or polyurethane-foam-
   lined box
Water

CAUTION: Sulfuric Acid is dangerous!

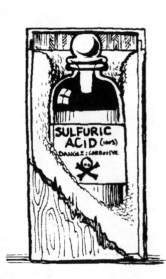 It is very strong, can cause severe burns, and even weak solutions can cause fingers to turn yellow. Water must never be poured into Sulfuric Acid because it will boil, spatter, and throw acid

all about. Sulfuric Acid should be used only under the supervision of a qualified chemist and when wearing safety glasses, long rubber gloves, rubber lab apron, and clothing that covers as much of the skin as possible. A face mask provides an extra measure of safety.

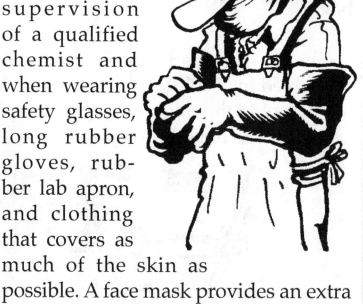

CAUTION: A glass of baking soda (Sodium Bicarbonate) solution should be kept on hand. The solution should be a mix of as much baking soda as will dissolve in water. If any acid get on the skin, the baking soda solution is dashed immediately on the skin.

CAUTION: The acid should be kept in a tightly sealed reagent bottle in a Styrofoam- or polyurethane foam-lined box.

The Purified Marijuana Extract is dissolved in absolute ethanol or pure methanol in the ratio of one gram extract to ten grams solvent.

CAUTION: There must be no water in this solution because of the way water reacts with Sulfuric Acid, which is added in the next step.

The next step is the addition of one drop of 100% Sulfuric Acid per gram of extract. The acid is added slowly, drop by drop, while the solution being created is stirred slowly and thoroughly with a long glass stirring rod.

The extract-alcohol-acid solution is put into a heat-proof glass pot. Pyrex is a common brand. The glass pot is put into the refluxing apparatus instead of the stainless steel pot. The heat-proof glass pot is used instead of the stainless steel pot because of the highly reactive nature of the acid, which attacks metal.

A two-hour reflux is then performed. The acid will not evaporate and will remain in the heat-proof glass pot.

The solution is then cooled.

The cooled solution and an equal volume of water plus 1/2 its volume of petroleum ether is poured into the petroleum ether extraction apparatus.

The petroleum ether extraction procedure is followed as before.

The mixture is poured into four volumes of water in a jug and gently inverted 25 times, letting off pressure between inversions.

The product is allowed to settle into separate layers, the petroleum ether-extract layer is blown off, and the water is discarded. This leaves a petroleum ether-extract-acid mix from which the acid must be purged.

To accomplish this, the petroleum ether-extract solution is poured into four times its volume of 5% sodium bicarbonate solution, which can be made by mixing baking soda and water in a ratio of one gram of baking

soda to twenty grams of water. This will neutralize the acid and leave a solution of Sodium Sulfate.

The mix is permitted to settle into layers. Then the petroleum ether-extract layer is blown off.

This step is repeated with pure water rather than baking soda solution.

Using the stainless steel pot in the refluxing apparatus and the solvent-collection pan in the colander, the extract and the petroleum ether are separated, and the petroleum ether is stored for future work.

It's Super Extract!

The extract contains a much higher percentage of THC, as determined by the amount of cannabidiol present. And that THC has been enhanced and is now the high-rotating isomeric form. All toxins have been removed.

Basting With Super-Extract

Supplies:
Flowering tops in good condition
Cheesecloth bag
Reflux apparatus
Sulfuric Acid
Supplies for isomerization
Basting syringe

Sophisticated marijuana users generally prefer discreet flowering tops in good, undamaged condition. When available, these can be intensified to hash-like strength.

A quantity of these flowering tops is selected and sewn into a cheesecloth bag. These are placed in the colander and refluxed in the reflux apparatus to remove their resins.

The extract is then purified by petroleum ether-extraction.

Next, the extract is isomerized with Sulfuric Acid, as previously described.

The resultant Super-Extract is dissolved into just as much ethanol as will be necessary to wet all of the deactivated flowers.

The ethanol/super-extract solution is drawn into a basting syringe and the dried, deactivated flowers are wetted with solution from the syringe.

Flowers soak up a lot of extract when dry, so if they are dried gently, being left out in air after applying the solution, then they will be able to soak up more extract from another subsequent application.

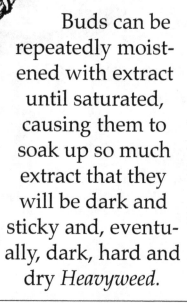

Buds can be repeatedly moistened with extract until saturated, causing them to soak up so much extract that they will be dark and sticky and, eventually, dark, hard and dry *Heavyweed*.

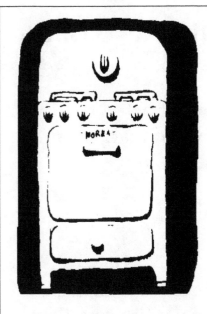

After the saturated flowers have been allowed to dry nearly completely in the atmosphere, some chemists complete the drying by placing the buds in a warm oven preheated to 200 degrees F or 93 degrees C.

CAUTION: The oven must turned off before placing the buds in it because they will exude explosive alcohol fumes.

Making "Artificial" Hashish

Supplies:
Isomerized Super-Extract
Powdered deactivated marijuana
Mortar and pestle
Wax paper
Rolling pin
Humidor

The making of hashish is an ancient art with countless variations. Many colors, flavors, consistencies, potencies and different methods of preparation exist all over the world, and since most of these are made possible only by local conditions, it's impractical to try and duplicate them.

AIN'T NOTHIN' BETTER'N A HASH SANDWICH!

"Artificial" hash can be made easily with home-made materials, and if done correctly will be indistinguishable from (and in some cases superior to) imported hashish.

A volume of pure *Isomerized Super-Extract* is heated it until it flows smoothly and then poured into a mortar with dried, powdered de-activated marijuana material. The two are ground together until the mixture is homogenous, and then more powder is added until the desired texture is arrived at.

The mixture is kneaded with the fingers and rolled out on a sheet of waxed paper.

The dough can be rolled into balls, eggs, hot-dog like cylinders, long, thin, worm-like pieces or other shapes, or rolled out beneath a rolling pin into wafers, slabs, or thin sheets.

The dough can also be rolled into many tiny balls which are then rolled in the dried powder and loosely forced together like a snowball.

The prepared material is put aside in a cool, slightly moist humidor to age. The medicine produced will be dark and spongy, and has superlative burning qualities due to its porosity. A similar hashish can be made by piling many thin layers of hash upon one another, separated by layers of marijuana dust.

Making Hash Oil

Supplies:
Petroleum ether separated extract
Norit (granulated activated charcoal)
Ethanol
Reflux apparatus
Fine filter paper funnel

Marijuana extract that looks like honey can be produced by removing the colored impurities from the extract which has been previously petroleum ether separated. This is done by dissolving the extract in ten times its volume of pure ethanol and mixing in a quantity of Norit (granulated activated charcoal) equal to half the weight of the extract.

Next, the mixture is filtered through fine filter paper in a funnel, and the ethanol removed by evaporation in the reflux apparatus. The residue is a thick fluid resembling dark amber honey, which may be smoked, eaten, or Isomerized.

Many people report good results from painting hash oil cigarette papers before rolling a joint.

Another technique is to dip joints into a tincture of extract in alcohol and dry before smoking.

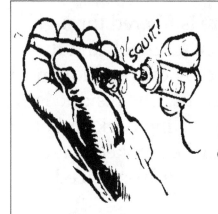

Some people inject oil into their joints with a small basting syringe or other appropriate device.

For smoking, little cup-shaped hollow may be may be shaped in a sheet of tin foil and a dab of hash oil can be placed at the bottom of the hollow. The dab can be heated from underneath with a match, lighter, or candle until it begins to emit smoke, which can be inhaled through a straw or tube.

For eating, hash oil is mixed with
butter, ground nuts, or some-
thing at least slightly oily
to aid in assimilating it
into the
system.

Conclusion

There are many ways to increase the
potency of marijuana for use. The
resulting potent preparations can be
smoked or used in
marijuana-containing foods.

The main advantages of potency increasing techniques are for those who smoke their marijuana. The more potent the preparation, the less need be smoked to achieve the desired level of effect. In this way, the various health hazards and irritations associated with smoking can be minimized.

Once again, the processes described in this book involve the use of volatile solvents and Sulfuric Acid, which are dangerous chemicals. Anyone interested in the application of the processes described in this book should work under the supervision of a qualified chemist and closely heed the warnings.

Now it's time for Dr. Atomic to sign off. Wishing all marijuana users the best of buds!

References

Adams, Roger, Ph. D., "Marijuana," *Bulletin New York Academy of Medicine*, vol. 18, 1942, *Marijuana: Medical Papers* edited by Tod. H. Mikuriya, M. D.

Boire, Richard Glen, *Marijuana Law, 2nd Edition*, Ronin, 1996.

Cervantes, Jorge, Robert Connell Clarke, and Ed Rosenthal, *Indoor Marijuana Horticulture*, Revised, 1993.

Cherniak, Laurence, *The Great Books of Hashish: Volume I, Book I: Morocco, Lebanon, Afghanistan, the Himalayas*, And/Or, 1979.

Cherniak, Laurence, *The Great Books of Hashish: Volume I, Book II: Marijuana Around The World, Sinsemilla, Stash, Opium*, Cherniak/Damele Publishing, 1982.

Clarke, Robert Connell, *Marijuana Botany: The Propagation and Breeding of Distinctive Cannabis*, Ronin., 1981.

Conrad, Chris, *Hemp for Health: The and Nutritional Uses of Cannabis Sativa* , Healing Arts, 1997.

Conrad, Chris, *Hemp: Lifeline to the Future*, Creative Xpressions, 1993.

Drake, Bill, *Marijuana: The Cultivator's Handbook*, Ronin, 1986.

Flowers, Tom, *Marijuana Flower Forcing: Secrets of Designer Growing*, Flowers Publishing, 1997.

Flowers, Tom, *Marijuana Herbal Cookbook: Recipes for Recreation and Health*, Flowers Publishing, 1995.

Formukong, E. A., A. T. Evans, and F. J. Evans, "Analgesic and Antiinflammatory Activity of Constitutents of Cannabis Sativa L."*Inflammation*, Vol. 12, No. 4, 1988.

Gieringer, Dale, Ph.D., "Marijuana Waterpipe/Vaporizer Study," *MAPS*, Spring 1996.

Gieringer, Dale, Ph.D., "Why Marijuana Smoke Harm Reduction?," *MAPS*, Spring 1996.

Gold, D., *Cannabis Alchemy*, Ronin, 1989.

Gottlieb, Adam, *The Art and Science of Cooking with Cannabis: The Most Effective Methods of Preparing Food and Drink with Marijuana, Hashish and Hash Oil*, Ronin, 1993.

Grinspoon, Lester, M.D., and James B. Bakalar, *Marihuana: The Forbidden Medicine*, Yale University Press,1993.

Herer, Jack, *The Emperor Wears No Clothes: The Authoritative Historical Record of the Cannabis Plant, Hemp Prohibition, and How Marijuana Can Still Save the World*, Hemp/Queen of Clubs Publishing, 1990.

Ludlow, Fitz Hugh, *The Hasheesh Eater*, City Lights, 1984.

Mathre, Mary Lynn, ed., *Cannabis in Medical Practice: A Legal, Historical, and Pharmacological Overview of the Therapeutic Uses of Marijuana*, McFarland, 1997.

Maurer, M., V. Henn, A. Dittrich, and A. Hofmann, "Delta-9-tetrahydrocannabinol Shows Antispastic and Analgesic Effects in a Single Case Double-blind Trial," *European Archives of Psychiatry and Clinical Neuroscience* 240 (1990).

Mechoulam, Raphael, ed., *Cannabinoids as Therapeutic Agents*, CRC, 1986.

Mikuriya, Tod H., M. D., ed., *Marijuana: Medical Papers 1839-1972*, Medi-Comp, 1973.

Murphy, Laura and Andrzej Bartke, eds., *Marijuana/Cannabinoids: Neurobiology and Neurophysiology*, CRC, 1992.

National Academy of Sciences, *Marijuana and Health*, National Academy, 1982.

Potter, Beverly A., and Dan Joy, *The Healing Magic Of Cannabis*, Ronin, 1998.

Randall, R.C., ed., *Cancer Treatment and Marijuana Therapy* ("Marijuana, Medicine, and the Law" series), Galen, 1990.

Randall, R.C., Editor, *Marijuana, Medicine, and the Law*, Galen, 1988.

Randall, R.C., Editor, *Marijuana, Medicine, and the Law: Volume II*, Galen, 1989.

Randall, R.C., *Marijuana and AIDS: Pot, Politics, and PWAs in America*, Galen, 1991.

Rathbun, Mary, and Dennis Peron, *Brownie Mary's Cookbook and Dennis Peron's Recipe for Social Change*, Trail of Smoke, 1996.

Robinson, Rowan, *The Great Book of Hemp: The Complete Guide to the Environmental, Commercial, and Uses of the World's Most Extraordinary Plant*, Park Street Press, 1996.

Rosenthal, Ed, Dale Gieringer, and Tod Mikuriya, M.D., *Marijuana Medical Handbook: A Guide to Therapeutic Use*, Quick American Archives, 1997.

Rosenthal, Ed, *Marijuana Growing Tips*, Quick American Trading, 1986.

Rosenthal, Ed, William Logan, and Jeffrey Steinborn, *Marijuana, the Law, and You*, Quick American Archives, 1995.

Shulgin, Alexander T., Ph.D., *Controlled Substances: Chemical and Legal Guide to Federal Drug Laws*, Ronin, 1992.

Stafford, Peter, *Psychedelics Encyclopedia*, Third Expanded Edition, Ronin, 1992.

Starks, Michael, *Marijuana Chemistry: Genetics, Processing, and Potency*, Ronin, 1990.

Tashkin, D. P., B. J. Shapiro, and I. A. Frank, "Acute Pulmonary Physiologic Effects of Smoked Marihuana and Oral Delta-9-tetrahydrocannabinol in Healthy Young Men," *New England Journal of Medicine* 289 (1973).

Tashkin, Donald P., M.D., Michael Simmons, and Virginia Clark, Ph.D., "Effect of Habitual Smoking of Marijuana Alone and with Tobacco on Nonspecific Airways Hyperreactivity," *Journal of Psychoactive Drugs*, Jan-Mar 1988.

Zuardi, A. W., I. Shirakawa, E. Finkelbarb, and I. G. Karniol, "Action of Cannabidiol on the Anxiety and Other Effects Produced by Delta-9-THC in Normal Subjects," *Psychopharmacology* 76 (1976).